Table of Contents

PRESIDENT OBAMA'S CLIMATE ACTION PLAN

"We, the people, still believe that our obligations as Americans are not just to ourselves, but to all posterity. We will respond to the threat of climate change, knowing that the failure to do so would betray our children and future generations. Some may still deny the overwhelming judgment of science, but none can avoid the devastating impact of raging fires and crippling drought and more powerful storms.

The path towards sustainable energy sources will be long and sometimes difficult. But America cannot resist this transition, we must lead it. We cannot cede to other nations the technology that will power new jobs and new industries, we must claim its promise. That's how we will maintain our economic vitality and our national treasure -- our forests and waterways, our croplands and snow-capped peaks. That is how we will preserve our planet, commanded to our care by God. That's what will lend meaning to the creed our fathers once declared."

-- President Obama, Second Inaugural Address, January 2013

THE CASE FOR ACTION

While no single step can reverse the effects of climate change, we have a moral obligation to future generations to leave them a planet that is not polluted and damaged. Through steady, responsible action to cut carbon pollution, we can protect our children's health and begin to slow the effects of climate change so that we leave behind a cleaner, more stable environment.

In 2009, President Obama made a pledge that by 2020, America would reduce its greenhouse gas emissions in the range of 17 percent below 2005 levels if all other major economies agreed to limit their emissions as well. Today, the President remains firmly committed to that goal and to building on the progress of his first term to help put us and the world on a sustainable long-term trajectory. Thanks in part to the Administration's success in doubling America's use of wind, solar, and geothermal energy and in establishing the toughest fuel economy standards in our history, we are creating new jobs, building new industries, and reducing dangerous carbon pollution which contributes to climate change. In fact, last year, carbon emissions from the energy sector fell to the lowest level in two decades. At the same time, while there is more work to do, we are more energy secure than at any time in recent history. In 2012, America's net oil imports fell to the lowest level in 20 years and we have become the world's leading producer of natural gas – the cleanest-burning fossil fuel.

While this progress is encouraging, climate change is no longer a distant threat – we are already feeling its impacts across the country and the world. Last year was the warmest year ever in the contiguous United States and about one-third of all Americans experienced 10 days or more of 100-degree heat. The 12 hottest years on record have all come in the last 15 years. Asthma rates have doubled in the past 30 years and our children will suffer more asthma attacks as air pollution gets worse. And increasing floods, heat waves, and droughts have put farmers out of business, which is already raising food prices dramatically.

These changes come with far-reaching consequences and real economic costs. Last year alone, there were 11 different weather and climate disaster events with estimated losses exceeding $1 billion each across the United States. Taken together, these 11 events resulted in over $110 billion in estimated damages, which would make it the second-costliest year on record.

In short, America stands at a critical juncture. Today, President Obama is putting forward a broad-based plan to cut the carbon pollution that causes climate change and affects public health. Cutting carbon pollution will help spark business innovation to modernize our power plants, resulting in cleaner forms of American-made energy that will create good jobs and cut our dependence on foreign oil. Combined with the Administration's other actions to increase the efficiency of our cars and household appliances, the President's plan will reduce the amount of energy consumed by American families, cutting down on their gas and utility bills. The plan, which consists of a wide variety of executive actions, has three key pillars:

1) **Cut Carbon Pollution in America:** In 2012, U.S. carbon emissions fell to the lowest level in two decades even as the economy continued to grow. To build on this progress, the Obama Administration is putting in place tough new rules to cut carbon pollution – just like we have for other toxins like mercury and arsenic – so we protect the health of our children and move our economy toward American-made clean energy sources that will create good jobs and lower home energy bills.

2) **Prepare the United States for the Impacts of Climate Change:** Even as we take new steps to reduce carbon pollution, we must also prepare for the impacts of a changing climate that are already being felt across the country. Moving forward, the Obama Administration will help state and local governments strengthen our roads, bridges, and shorelines so we can better protect people's homes, businesses and way of life from severe weather.

3) **Lead International Efforts to Combat Global Climate Change and Prepare for its Impacts:** Just as no country is immune from the impacts of climate change, no country can meet this challenge alone. That is why it is imperative for the United States to couple action at home with leadership internationally. America must help forge a truly global solution to this global challenge by galvanizing international action to significantly reduce emissions (particularly among the major emitting countries), prepare for climate impacts, and drive progress through the international negotiations.

Climate change represents one of our greatest challenges of our time, but it is a challenge uniquely suited to America's strengths. Our scientists will design new fuels, and our farmers will grow them. Our engineers to devise new sources of energy, our workers will build them, and our businesses will sell them. All of us will need to do our part. If we embrace this challenge, we will not just create new jobs and new industries and keep America on the cutting edge; we will save lives, protect and preserve our treasured natural resources, cities, and coastlines for future generations.

What follows is a blueprint for steady, responsible national and international action to slow the effects of climate change so we leave a cleaner, more stable environment for future generations. It highlights progress already set in motion by the Obama Administration to advance these goals and sets forth new steps to achieve them.

In 2009, President Obama made a commitment to reduce U.S. greenhouse gas emissions in the range of 17 percent below 2005 levels by 2020. The President remains firmly committed to achieving that goal. While there is more work to do, the Obama Administration has already made significant progress by doubling generation of electricity from wind, solar, and geothermal, and by establishing historic new fuel economy standards. Building on these achievements, this document outlines additional steps the Administration will take – in partnership with states, local communities, and the private sector – to continue on a path to meeting the President's 2020 goal.

I.   *Deploying Clean Energy*

Cutting Carbon Pollution from Power Plants: Power plants are the largest concentrated source of emissions in the United States, together accounting for roughly one-third of all domestic greenhouse gas emissions. We have already set limits for arsenic, mercury, and lead, but there is no federal rule to prevent power plants from releasing as much carbon pollution as they want. Many states, local governments, and companies have taken steps to move to cleaner electricity sources. More than 35 states have renewable energy targets in place, and more than 25 have set energy efficiency targets.

Despite this progress at the state level, there are no federal standards in place to reduce carbon pollution from power plants. In April 2012, as part of a continued effort to modernize our electric power sector, the Obama Administration proposed a carbon pollution standard for new power plants. The Environmental Protection Agency's proposal reflects and reinforces the ongoing trend towards cleaner technologies, with natural gas increasing its share of electricity generation in recent years, principally through market forces and renewables deployment growing rapidly to account for roughly half of new generation capacity installed in 2012.

With abundant clean energy solutions available, and building on the leadership of states and local governments, we can make continued progress in reducing power plant pollution to improve public health and the environment while supplying the reliable, affordable power needed for economic growth. By doing so, we will continue to drive American leadership in clean energy technologies, such as efficient natural gas, nuclear, renewables, and clean coal technology.

To accomplish these goals, President Obama is issuing a Presidential Memorandum directing the Environmental Protection Agency to work expeditiously to complete carbon pollution standards for both new and existing power plants. This work will build on the successful first-term effort to develop greenhouse gas and fuel economy standards for cars and trucks. In developing the standards, the President has asked the Environmental Protection Agency to build on state leadership, provide flexibility, and take advantage of a wide range of energy sources and technologies including many actions in this plan.

Promoting American Leadership in Renewable Energy: During the President's first term, the United States more than doubled generation of electricity from wind, solar, and geothermal sources. To ensure America's continued leadership position in clean energy, President Obama has set a goal to double renewable electricity generation once again by 2020. In order to meet this ambitious target, the Administration is announcing a number of new efforts in the following key areas:

- **Accelerating Clean Energy Permitting:** In 2012 the President set a goal to issue permits for 10 gigawatts of renewables on public lands by the end of the year. The Department of the Interior achieved this goal ahead of schedule and the President has directed it to permit an additional 10 gigawatts by 2020. Since 2009, the Department of Interior has approved 25 utility-scale solar facilities, nine wind farms, and 11 geothermal plants, which will provide enough electricity to power 4.4 million homes and support an estimated 17,000 jobs. The Administration is also taking steps to encourage the development of hydroelectric power at existing dams. To develop and demonstrate improved permitting procedures for such projects, the Administration will designate the Red Rock Hydroelectric Plant on the Des Moines River in Iowa to participate in its Infrastructure Permitting Dashboard for high-priority projects. Also, the Department of Defense – the single largest consumer of energy in the United States – is committed to deploying 3 gigawatts of renewable energy on military installations, including solar, wind, biomass, and geothermal, by 2025. In addition, federal agencies are setting a new goal of reaching 100 megawatts of installed renewable capacity across the federally subsidized housing stock by 2020. This effort will include conducting a survey of current projects in order to track progress and facilitate the sharing of best practices.

- **Expanding and Modernizing the Electric Grid:** Upgrading the country's electric grid is critical to our efforts to make electricity more reliable, save consumers money on their energy bills, and promote clean energy sources. To advance these important goals, President Obama signed a Presidential Memorandum this month that directs federal agencies to streamline the siting, permitting and review process for transmission projects across federal, state, and tribal governments.

Unlocking Long-Term Investment in Clean Energy Innovation: The Fiscal Year 2014 Budget continues the President's commitment to keeping the United States at the forefront of clean energy research, development, and deployment by increasing funding for clean energy technology across all agencies by 30 percent, to approximately $7.9 billion. This includes investment in a range of energy technologies, from advanced biofuels and emerging nuclear technologies – including small modular reactors – to clean coal. To continue America's leadership in clean energy innovation, the Administration will also take the following steps:

- **Spurring Investment in Advanced Fossil Energy Projects:** In the coming weeks, the Department of Energy will issue a Federal Register Notice announcing a draft of a solicitation that would make up to $8 billion in (self-pay) loan guarantee authority available for a wide array of advanced fossil energy projects under its Section 1703 loan guarantee program. This solicitation is designed to support investments in innovative technologies that can cost-effectively meet financial and policy goals, including the avoidance, reduction, or sequestration of anthropogenic emissions of greenhouse gases. The proposed solicitation will cover a broad range of advanced fossil energy projects. Reflecting the Department's commitment to continuous improvement in program management, it will take comment on the draft solicitation, with a plan to issue a final solicitation by the fall of 2013.

- **Instituting a Federal Quadrennial Energy Review:** Innovation and new sources of domestic energy supply are transforming the nation's energy marketplace, creating economic opportunities at the same time they raise environmental challenges. To ensure that federal energy policy meets our economic, environmental, and security goals in this changing landscape, the Administration will conduct a Quadrennial Energy Review which will be led by the White House Domestic Policy Council and Office of Science and Technology Policy, supported by a Secretariat established at the Department of Energy, and involving the robust engagement of federal agencies and outside stakeholders. This first-ever review will focus on infrastructure challenges, and will identify the threats, risks, and opportunities for U.S. energy and climate security, enabling the federal government to translate policy goals into a set of analytically based, clearly articulated, sequenced and integrated actions, and proposed investments over a four-year planning horizon.

II.   Building a 21st-Century Transportation Sector

Increasing Fuel Economy Standards: Heavy-duty vehicles are currently the second largest source of greenhouse gas emissions within the transportation sector. In 2011, the Obama Administration finalized the first-ever fuel economy standards for Model Year 2014-2018 for heavy-duty trucks, buses, and vans. These standards will reduce greenhouse gas emissions by approximately 270 million metric tons and save 530 million barrels of oil. During the President's second term, the Administration will once again partner with industry leaders and other key stakeholders to develop post-2018 fuel economy standards for heavy-duty vehicles to further reduce fuel consumption through the application of advanced cost-effective technologies and continue efforts to improve the efficiency of moving goods across the United States.

The Obama Administration has already established the toughest fuel economy standards for passenger vehicles in U.S. history. These standards require an average performance equivalent of 54.5 miles per gallon by 2025, which will save the average driver more than $8,000 in fuel costs over the lifetime of the vehicle and eliminate six billion metric tons of carbon pollution – more than the United States emits in an entire year.

Developing and Deploying Advanced Transportation Technologies: Biofuels have an important role to play in increasing our energy security, fostering rural economic development, and reducing greenhouse gas emissions from the transportation sector. That is why the Administration supports the Renewable Fuels Standard, and is investing in research and development to help bring next-generation biofuels on line. For example, the United States Navy and Departments of Energy and Agriculture are working with the private sector to accelerate the development of cost-competitive advanced biofuels for use by the military and commercial sectors. More broadly, the Administration will continue to leverage partnerships between the private and public sectors to deploy cleaner fuels, including advanced batteries and fuel cell technologies, in every transportation mode. The Department of Energy's eGallon informs drivers about electric car operating costs in their state – the national average is only $1.14 per gallon of gasoline equivalent, showing the promise for consumer pocketbooks of electric-powered vehicles. In addition, in the coming months, the Department of Transportation will work with other agencies to further explore strategies for integrating alternative fuel vessels into the U.S. flag fleet. Further, the Administration will continue to work with states, cities and towns through the Department of Transportation, the Department of Housing and Urban Development, and the Environmental Protection Agency to improve transportation options, and lower transportation costs while protecting the environment in communities nationwide.

III.   Cutting Energy Waste in Homes, Businesses, and Factories

Reducing Energy Bills for American Families and Businesses: Energy efficiency is one of the clearest and most cost-effective opportunities to save families money, make our businesses more competitive, and reduce greenhouse gas emissions. In the President's first term, the Department of Energy and the Department of Housing and Urban Development completed efficiency upgrades in more than one million homes, saving many families more than $400 on their heating and cooling bills in the first year alone. The Administration will take a range of new steps geared towards achieving President Obama's goal of doubling energy productivity by 2030 relative to 2010 levels:

- **Establishing a New Goal for Energy Efficiency Standards:** In President Obama's first term, the Department of Energy established new minimum efficiency standards for dishwashers, refrigerators, and many other products. Through 2030, these standards will cut consumers' electricity bills by hundreds of billions of dollars and save enough electricity to power more than 85 million homes for two years. To build on this success, the Administration is setting a new goal: Efficiency standards for appliances and federal buildings set in the first and second terms combined will reduce carbon pollution by at least 3 billion metric tons cumulatively by 2030 – equivalent to nearly one-half of the carbon pollution from the entire U.S. energy sector for one year – while continuing to cut families' energy bills.

- **Reducing Barriers to Investment in Energy Efficiency:** Energy efficiency upgrades bring significant cost savings, but upfront costs act as a barrier to more widespread investment. In response, the Administration is committing to a number of new executive actions. As soon as this fall, the Department of Agriculture's Rural Utilities Service will finalize a proposed update to its Energy Efficiency and Conservation Loan Program to provide up to $250 million for rural utilities to finance efficiency investments by businesses and homeowners across rural America. The Department is also streamlining its Rural Energy for America program to provide grants and loan guarantees directly to agricultural producers and rural small businesses for energy efficiency and renewable energy systems.

  In addition, the Department of Housing and Urban Development's efforts include a $23 million Multifamily Energy Innovation Fund designed to enable affordable housing providers, technology firms, academic institutions, and philanthropic organizations to test new approaches to deliver cost-effective residential energy. In order to advance ongoing efforts and bring stakeholders together, the Federal Housing Administration will convene representatives of the lending community and other key stakeholders for a mortgage roundtable in July to identify options for factoring energy efficiency into the mortgage underwriting and appraisal process upon sale or refinancing of new or existing homes.

- **Expanding the President's Better Buildings Challenge:** The Better Buildings Challenge, focused on helping American commercial and industrial buildings become at least 20 percent more energy efficient by 2020, is already showing results. More than 120 diverse organizations, representing over 2 billion square feet are on track to meet the 2020 goal: cutting energy use by an average 2.5 percent annually, equivalent to about $58 million in energy savings per year. To continue this success, the Administration will expand the program to multifamily housing – partnering both with private and affordable building owners and public housing agencies to cut energy waste. In addition, the Administration is launching the Better Buildings Accelerators, a new track that will support and encourage adoption of State and local policies to cut energy waste, building on the momentum of ongoing efforts at that level.

IV.   *Reducing Other Greenhouse Gas Emissions*

Curbing Emissions of Hydrofluorocarbons: Hydrofluorocarbons (HFCs), which are primarily used for refrigeration and air conditioning, are potent greenhouse gases. In the United States, emissions of HFCs are expected to nearly triple by 2030, and double from current levels of 1.5 percent of greenhouse gas emissions to 3 percent by 2020.

To reduce emissions of HFCs, the United States can and will lead both through international diplomacy as well as domestic actions. In fact, the Administration has already acted by including a flexible and powerful incentive in the fuel economy and carbon pollution standards for cars and trucks to encourage automakers to reduce HFC leakage and transition away from the most potent HFCs in vehicle air conditioning systems. Moving forward, the Environmental Protection Agency will use its authority through the Significant New Alternatives Policy Program to encourage private sector investment in low-emissions technology by identifying and approving climate-friendly chemicals while prohibiting certain uses of the most harmful chemical alternatives. In addition, the President has directed his Administration to purchase cleaner alternatives to HFCs whenever feasible and transition over time to equipment that uses safer and more sustainable alternatives.

Reducing Methane Emissions: Curbing emissions of methane is critical to our overall effort to address global climate change. Methane currently accounts for roughly 9 percent of domestic greenhouse gas emissions and has a global warming potential that is more than 20 times greater than carbon dioxide. Notably, since 1990, methane emissions in the United States have decreased by 8 percent. This has occurred in part through partnerships with industry, both at home and abroad, in which we have demonstrated that we have the technology to deliver emissions reductions that benefit both our economy and the environment. To achieve additional progress, the Administration will:

- **Developing an Interagency Methane Strategy:** The Environmental Protection Agency and the Departments of Agriculture, Energy, Interior, Labor, and Transportation will develop a comprehensive, interagency methane strategy. The group will focus on assessing current emissions data, addressing data gaps, identifying technologies and best practices for reducing emissions, and identifying existing authorities and incentive-based opportunities to reduce methane emissions.

- **Pursuing a Collaborative Approach to Reducing Emissions:** Across the economy, there are multiple sectors in which methane emissions can be reduced, from coal mines and landfills to agriculture and oil and gas development. For example, in the agricultural sector, over the last three years, the Environmental Protection Agency and the Department of Agriculture have worked with the dairy industry to increase the adoption of methane digesters through loans, incentives, and other assistance. In addition, when it comes to the oil and gas sector, investments to build and upgrade gas pipelines will not only put more Americans to work, but also reduce emissions and enhance economic productivity. For example, as part of the Administration's effort to improve federal permitting for infrastructure projects, the interagency Bakken Federal Executive Group is working with industry, as well as state and tribal agencies, to advance the production of oil and gas in the Bakken while helping to reduce venting and flaring. Moving forward, as part of the effort to develop an interagency methane strategy, the Obama Administration will work collaboratively with state governments, as well as the private sector, to reduce emissions across multiple sectors, improve air quality, and achieve public health and economic benefits.

Preserving the Role of Forests in Mitigating Climate Change: America's forests play a critical role in addressing carbon pollution, removing nearly 12 percent of total U.S. greenhouse gas emissions each year. In the face of a changing climate and increased risk of wildfire, drought, and pests, the capacity of our forests to absorb carbon is diminishing. Pressures to develop forest lands for urban or agricultural uses also contribute to the decline of forest carbon sequestration. Conservation and sustainable management can help to ensure our forests continue to remove carbon from the atmosphere while also improving soil and water quality, reducing wildfire risk, and otherwise managing forests to be more resilient in the fact of climate change. The Administration is working to identify new approaches to protect and restore our forests, as well as other critical landscapes including grasslands and wetlands, in the face of a changing climate.

V. *Leading at the Federal Level*

Leading in Clean Energy: President Obama believes that the federal government must be a leader in clean energy and energy efficiency. Under the Obama Administration, federal agencies have reduced greenhouse gas emissions by more than 15 percent – the equivalent of permanently taking 1.5 million cars off the road. To build on this record, the Administration is establishing a new goal: The federal government will consume 20 percent of its electricity from renewable sources by 2020 – more than double the current goal of 7.5 percent. In addition, the federal government will continue to pursue greater energy efficiency that reduces greenhouse gas emissions and saves taxpayer dollars.

Federal Government Leadership in Energy Efficiency: On December 2, 2011, President Obama signed a memorandum entitled "Implementation of Energy Savings Projects and Performance-Based Contracting for Energy Savings," challenging federal agencies, in support of the Better Buildings Challenge, to enter into $2 billion worth of performance-based contracts within two years. Performance contracts drive economic development, utilize private sector innovation, and increase efficiency at minimum costs to the taxpayer, while also providing long-term savings in energy costs. Federal agencies have committed to a pipeline of nearly $2.3 billion from over 300 reported projects. In coming months, the Administration will take a number of actions to strengthen efforts to promote energy efficiency, including through performance contracting. For example, in order to increase access to capital markets for investments in energy efficiency, the Administration will initiate a partnership with the private sector to work towards a standardized contract to finance federal investments in energy efficiency. Going forward, agencies will also work together to synchronize building codes – leveraging those policies to improve the efficiency of federally owned and supported building stock. Finally, the Administration will leverage the "Green Button" standard – which aggregates energy data in a secure, easy to use format – within federal facilities to increase their ability to manage energy consumption, reduce greenhouse gas emissions, and meet sustainability goals.

As we act to curb the greenhouse gas pollution that is driving climate change, we must also prepare for the impacts that are too late to avoid. Across America, states, cities, and communities are taking steps to protect themselves by updating building codes, adjusting the way they manage natural resources, investing in more resilient infrastructure, and planning for rapid recovery from damages that nonetheless occur. The federal government has an important role to play in supporting community-based preparedness and resilience efforts, establishing policies that promote preparedness, protecting critical infrastructure and public resources, supporting science and research germane to preparedness and resilience, and ensuring that federal operations and facilities continue to protect and serve citizens in a changing climate.

The Obama Administration has been working to strengthen America's climate resilience since its earliest days. Shortly after coming into office, President Obama established an Interagency Climate Change Adaptation Task Force and, in October 2009, the President signed an Executive Order directing it to recommend ways federal policies and programs can better prepare the Nation for change. In May 2010, the Task Force hosted the first National Climate Adaptation Summit, convening local and regional stakeholders and decision-makers to identify challenges and opportunities for collaborative action.

In February 2013, federal agencies released Climate Change Adaptation Plans for the first time, outlining strategies to protect their operations, missions, and programs from the effects of climate change. The Department of Transportation, for example, is developing guidance for incorporating climate change and extreme weather event considerations into coastal highway projects, and the Department of Homeland Security is evaluating the challenges of changing conditions in the Arctic and along our Nation's borders. Agencies have also partnered with communities through targeted grant and technical-assistance programs—for example, the Environmental Protection Agency is working with low-lying communities in North Carolina to assess the vulnerability of infrastructure investments to sea level rise and identify solutions to reduce risks. And the Administration has continued, through the U.S. Global Change Research Program, to support science and monitoring to expand our understanding of climate change and its impacts.

Going forward, the Administration will expand these efforts into three major, interrelated initiatives to better prepare America for the impacts of climate change:

I. *Building Stronger and Safer Communities and Infrastructure*

By necessity, many states, cities, and communities are already planning and preparing for the impacts of climate change. Hospitals must build capacity to serve patients during more frequent heat waves, and urban planners must plan for the severe storms that infrastructure will need to withstand. Promoting on-the-ground planning and resilient infrastructure will be at the core of our work to strengthen America's communities. Specific actions will include:

Directing Agencies to Support Climate-Resilient Investment: The President will direct federal agencies to identify and remove barriers to making climate-resilient investments; identify and remove counterproductive policies that increase vulnerabilities; and encourage and support smarter, more resilient investments, including through agency grants, technical assistance, and other programs, in sectors from transportation and water management to conservation and disaster relief. Agencies will also be directed to ensure that climate risk-management considerations are fully integrated into federal infrastructure and natural resource management planning. To begin meeting this challenge, the Environmental Protection Agency is committing to integrate considerations of climate change impacts and adaptive measures into major programs, including its Clean Water and Drinking Water State Revolving Funds and grants for brownfields cleanup, and the Department of Housing and Urban Development is already requiring grant recipients in the Hurricane Sandy–affected region to take sea-level rise into account.

Establishing a State, Local, and Tribal Leaders Task Force on Climate Preparedness: To help agencies meet the above directive and to enhance local efforts to protect communities, the President will establish a short-term task force of state, local, and tribal officials to advise on key actions the federal government can take to better support local preparedness and resilience-building efforts. The task force will provide recommendations on removing barriers to resilient investments, modernizing grant and loan programs to better support local efforts, and developing information and tools to better serve communities.

Supporting Communities as they Prepare for Climate Impacts: Federal agencies will continue to provide targeted support and assistance to help communities prepare for climate-change impacts. For example, throughout 2013, the Department of Transportation's Federal Highway Administration is working with 19 state and regional partners and other federal agencies to test approaches for assessing local transportation infrastructure vulnerability to climate change and extreme weather and for improving resilience. The Administration will continue to assist tribal communities on preparedness through the Bureau of Indian Affairs, including through pilot projects and by supporting participation in federal initiatives that assess climate change vulnerabilities and develop regional solutions. Through annual federal agency "Environmental Justice Progress Reports," the Administration will continue to identify innovative ways to help our most vulnerable communities prepare for and recover from the impacts of climate change. The importance of critical infrastructure independence was brought home in the Sandy response. The Federal Emergency Management Agency and the Department of Energy are working with the private sector to address simultaneous restoration of electricity and fuels supply.

Boosting the Resilience of Buildings and Infrastructure: The National Institute of Standards and Technology will convene a panel on disaster-resilience standards to develop a comprehensive, community-based resilience framework and provide guidelines for consistently safe buildings and infrastructure – products that can inform the development of private-sector standards and codes. In addition, building on federal agencies' "Climate Change Adaptation Plans," the Administration will continue efforts to increase the resilience of federal facilities and infrastructure. The Department of Defense, for example, is assessing the relative vulnerability of its coastal facilities to climate change. In addition, the President's FY 2014 Budget proposes $200 million through the Transportation Leadership Awards program for Climate Ready Infrastructure in communities that build enhanced preparedness into their planning efforts, and that have proposed or are ready to break ground on infrastructure projects, including transit and rail, to improve resilience.

Rebuilding and Learning from Hurricane Sandy: In August 2013, President Obama's Hurricane Sandy Rebuilding Task Force will deliver to the President a rebuilding strategy to be implemented in Sandy-affected regions and establishing precedents that can be followed elsewhere. The Task Force and federal agencies are also piloting new ways to support resilience in the Sandy-affected region; the Task Force, for example, is hosting a regional "Rebuilding by Design" competition to generate innovative solutions to enhance resilience. In the transportation sector, the Department of Transportation's Federal Transit Administration (FTA) is dedicating $5.7 billion to four of the area's most impacted transit agencies, of which $1.3 billion will be allocated to locally prioritized projects to make transit systems more resilient to future disasters. FTA will also develop a competitive process for additional funding to identify and support larger, stand-alone resilience projects in the impacted region. To build coastal resilience, the Department of the Interior will launch a $100 million competitive grant program to foster partnerships and promote resilient natural systems while enhancing green spaces and wildlife habitat near urban populations. An additional $250 million will be allocated to support projects for coastal restoration and resilience across the region. Finally, with partners, the U.S. Army Corps of Engineers is conducting a $20 million study to identify strategies to reduce the vulnerability of Sandy-affected coastal communities to future large-scale flood and storm events, and the National Oceanic and Atmospheric Administration will strengthen long-term coastal observations and provide technical assistance to coastal communities.

II.   *Protecting our Economy and Natural Resources*

Climate change is affecting nearly every aspect of our society, from agriculture and tourism to the health and safety of our citizens and natural resources. To help protect critical sectors, while also targeting hazards that cut across sectors and regions, the Administration will mount a set of sector- and hazard-specific efforts to protect our country's vital assets, to include:

Identifying Vulnerabilities of Key Sectors to Climate Change: The Department of Energy will soon release an assessment of climate-change impacts on the energy sector, including power-plant disruptions due to drought and the disruption of fuel supplies during severe storms, as well as potential opportunities to make our energy infrastructure more resilient to these risks. In 2013, the Department of Agriculture and Department of the Interior released several studies outlining the challenges a changing climate poses for America's agricultural enterprise, forests, water supply, wildlife, and public lands. This year and next, federal agencies will report on the impacts of climate change on other key sectors and strategies to address them, with priority efforts focusing on health, transportation, food supplies, oceans, and coastal communities.

Promoting Resilience in the Health Sector: The Department of Health and Human Services will launch an effort to create sustainable and resilient hospitals in the face of climate change. Through a public-private partnership with the healthcare industry, it will identify best practices and provide guidance on affordable measures to ensure that our medical system is resilient to climate impacts. It will also collaborate with partner agencies to share best practices among federal health facilities. And, building on lessons from pilot projects underway in 16 states, it will help train public-health professionals and community leaders to prepare their communities for the health consequences of climate change, including through effective communication of health risks and resilience measures.

Promoting Insurance Leadership for Climate Safety: Recognizing the critical role that the private sector plays in insuring assets and enabling rapid recovery after disasters, the Administration will convene representatives from the insurance industry and other stakeholders to explore best practices for private and public insurers to manage their own processes and investments to account for climate change risks and incentivize policy holders to take steps to reduce their exposure to these risks.

Conserving Land and Water Resources: America's ecosystems are critical to our nation's economy and the lives and health of our citizens. These natural resources can also help ameliorate the impacts of climate change, if they are properly protected. The Administration has invested significantly in conserving relevant ecosystems, including working with Gulf State partners after the Deepwater Horizon spill to enhance barrier islands and marshes that protect communities from severe storms. The Administration is also implementing climate-adaptation strategies that promote resilience in fish and wildlife populations, forests and other plant communities, freshwater resources, and the ocean. Building on these efforts, the President is also directing federal agencies to identify and evaluate additional approaches to improve our natural defenses against extreme weather, protect biodiversity and conserve natural resources in the face of a changing climate, and manage our public lands and natural systems to store more carbon.

Maintaining Agricultural Sustainability: Building on the existing network of federal climate-science research and action centers, the Department of Agriculture is creating seven new Regional Climate Hubs to deliver tailored, science-based knowledge to farmers, ranchers, and forest landowners. These hubs will work with universities and other partners, including the Department of the Interior and the National Oceanic and Atmospheric Administration, to support climate resilience. Its Natural Resources Conservation Service and the Department of the Interior's Bureau of Reclamation are also providing grants and technical support to agricultural water users for more water-efficient practices in the face of drought and long-term climate change.

Managing Drought: Leveraging the work of the National Disaster Recovery Framework for drought, the Administration will launch a cross-agency National Drought Resilience Partnership as a "front door" for communities seeking help to prepare for future droughts and reduce drought impacts. By linking information (monitoring, forecasts, outlooks, and early warnings) with drought preparedness and longer-term resilience strategies in critical sectors, this effort will help communities manage drought-related risks.

Reducing Wildfire Risks: With tribes, states, and local governments as partners, the Administration has worked to make landscapes more resistant to wildfires, which are exacerbated by heat and drought conditions resulting from climate change. Federal agencies will expand and prioritize forest and rangeland restoration efforts in order to make natural areas and communities less vulnerable to catastrophic fire. The Department of the Interior and Department of Agriculture, for example, are launching a Western Watershed Enhancement Partnership – a pilot effort in five western states to reduce wildfire risk by removing extra brush and other flammable vegetation around critical areas such as water reservoirs.

Preparing for Future Floods: To ensure that projects funded with taxpayer dollars last as long as intended, federal agencies will update their flood-risk reduction standards for federally funded projects to reflect a consistent approach that accounts for sea-level rise and other factors affecting flood risks. This effort will incorporate the most recent science on expected rates of sea-level rise (which vary by region) and build on work done by the Hurricane Sandy Rebuilding Task Force, which announced in April 2013 that all federally funded Sandy-related rebuilding projects must meet a consistent flood risk reduction standard that takes into account increased risk from extreme weather events, sea-level rise, and other impacts of climate change.

III.   *Using Sound Science to Manage Climate Impacts*

Scientific data and insights are essential to help government officials, communities, and businesses better understand and manage the risks associated with climate change. The Administration will continue to lead in advancing the science of climate measurement and adaptation and the development of tools for climate-relevant decision-making by focusing on increasing the availability, accessibility, and utility of relevant scientific tools and information. Specific actions will include:

Developing Actionable Climate Science: The President's Fiscal Year 2014 Budget provides more than $2.7 billion, largely through the 13-agency U.S. Global Change Research Program, to increase understanding of climate-change impacts, establish a public-private partnership to explore risk and catastrophe modeling, and develop the information and tools needed by decision-makers to respond to both long-term climate change impacts and near-term effects of extreme weather.

Assessing Climate-Change Impacts in the United States: In the spring of 2014, the Obama Administration will release the third U.S. National Climate Assessment, highlighting new advances in our understanding of climate-change impacts across all regions of the United States and on critical sectors of the economy, including transportation, energy, agriculture, and ecosystems and biodiversity. For the first time, the National Climate Assessment will focus not only on dissemination of scientific information but also on translating scientific insights into practical, useable knowledge that can help decision-makers anticipate and prepare for specific climate-change impacts.

Launching a Climate Data Initiative: Consistent with the President's May 2013 Executive Order on Open Data – and recognizing that freely available open government data can fuel entrepreneurship, innovation, scientific discovery, and public benefits – the Administration is launching a Climate Data Initiative to leverage extensive federal climate-relevant data to stimulate innovation and private-sector entrepreneurship in support of national climate-change preparedness.

Providing a Toolkit for Climate Resilience: Federal agencies will create a virtual climate-resilience toolkit that centralizes access to data-driven resilience tools, services, and best practices, including those developed through the Climate Data Initiative. The toolkit will provide easy access to existing resources as well as new tools, including: interactive sea-level rise maps and a sea-level-rise calculator to aid post-Sandy rebuilding in New York and New Jersey, new NOAA storm surge models and interactive maps from the National Oceanic and Atmospheric Administration that provide risk information by combining tidal data, projected sea levels and storm wave heights, a web-based tool that will allow developers to integrate NASA climate imagery into websites and mobile apps, access to the U.S. Geological Survey's "visualization tool" to assess the amount of carbon absorbed by landscapes, and a Stormwater Calculator and Climate Assessment Tool developed to help local governments assess stormwater-control measures under different precipitation and temperature scenarios.

The Obama Administration is working to build on the actions that it is taking domestically to achieve significant global greenhouse gas emission reductions and enhance climate preparedness through major international initiatives focused on spurring concrete action, including bilateral initiatives with China, India, and other major emitting countries. These initiatives not only serve to support the efforts of the United States and others to achieve our goals for 2020, but also will help us move beyond those and bend the post-2020 global emissions trajectory further. As a key part of this effort, we are also working intensively to forge global responses to climate change through a number of important international negotiations, including the United Nations Framework Convention on Climate Change.

I.   *Working with Other Countries to Take Action to Address Climate Change*

Enhancing Multilateral Engagement with Major Economies: In 2009, President Obama launched the Major Economies Forum on Energy and Climate, a high-level forum that brings together 17 countries that account for approximately 75 percent of global greenhouse gas emissions, in order to support the international climate negotiations and spur cooperative action to combat climate change. The Forum has been successful on both fronts – having contributed significantly to progress in the broader negotiations while also launching the Clean Energy Ministerial to catalyze the development and deployment of clean energy and efficiency solutions. We are proposing that the Forum build on these efforts by launching a major initiative this year focused on further accelerating efficiency gains in the buildings sector, which accounts for approximately one-third of global carbon pollutions from the energy sector.

Expanding Bilateral Cooperation with Major Emerging Economies:
From the outset, the Obama Administration has sought to intensify bilateral climate cooperation with key major emerging economies, through initiatives like the U.S.-China Clean Energy Research Center, the U.S.-India Partnership to Advance Clean Energy, and the Strategic Energy Dialogue with Brazil.

We will be building on these successes and finding new areas for cooperation in the second term, and we are already making progress: Just this month, President Obama and President Xi Jinping of China reached an historic agreement at their first summit to work to use the expertise and institutions of the Montreal Protocol to phase down the consumption and production of HFCs, a highly potent greenhouse gas. The impact of phasing out HFCs by 2050 would be equivalent to the elimination of two years' worth of greenhouse gas emissions from all sources.

Combatting Short-Lived Climate Pollutants: Pollutants such as methane, black carbon, and many HFCs are relatively short-lived in the atmosphere, but have more potent greenhouse effects than carbon dioxide. In February 2012, the United States launched the Climate and Clean Air Coalition to Reduce Short-Lived Climate Pollution, which has grown to include more than 30 country partners and other key partners such as the World Bank and the U.N. Environment Programme. Major efforts include reducing methane and black carbon from waste and landfills. We are also leading through the Global Methane Initiative, which works with 42 partner countries and an extensive network of over 1,100 private sector participants to reduce methane emissions.

Reducing Emissions from Deforestation and Forest Degradation: Greenhouse gas emissions from deforestation, agriculture, and other land use constitute approximately one-third of global emissions. In some developing countries, as much as 80 percent of these emissions come from the land sector. To meet this challenge, the Obama Administration is working with partner countries to put in place the systems and institutions necessary to significantly reduce global land-use-related emissions, creating new models for rural development that generate climate benefits, while conserving biodiversity, protecting watersheds, and improving livelihoods.

In 2012 alone, the U.S. Agency for International Development's bilateral and regional forestry programs contributed to reducing more than 140 million tons of carbon dioxide emissions, including through support for multilateral initiatives such as the Forest Investment Program and the Forest Carbon Partnership Facility. In Indonesia, the Millennium Challenge Corporation is funding a five-year "Green Prosperity" program that supports environmentally sustainable, low carbon economic development in select districts.

The Obama Administration is also working to address agriculture-driven deforestation through initiatives such as the Tropical Forest Alliance 2020, which brings together governments, the private sector, and civil society to reduce tropical deforestation related to key agricultural commodities, which we will build upon.

Expanding Clean Energy Use and Cut Energy Waste: Roughly 84 percent of current carbon dioxide emissions are energy-related and about 65 percent of all greenhouse gas emissions can be attributed to energy supply and energy use. The Obama Administration has promoted the expansion of renewable, clean, and efficient energy sources and technologies worldwide through:

- Financing and regulatory support for renewable and clean energy projects
- Actions to promote fuel switching from oil and coal to natural gas or renewables
- Support for the safe and secure use of nuclear power
- Cooperation on clean coal technologies
- Programs to improve and disseminate energy efficient technologies

In the past three years we have reached agreements with more than 20 countries around the world, including Mexico, South Africa, and Indonesia, to support low emission development strategies that help countries to identify the best ways to reduce greenhouse gas emissions while growing their economies. Among the many initiatives that we have launched are:

- The U.S. Africa Clean Energy Finance Initiative, which aligns grant-based assistance with project planning expertise from the U.S. Trade and Development Agency and financing and risk mitigation tools from the U.S. Overseas Private Investment Corporation to unlock up to $1 billion in clean energy financing.

- The U.S.-Asia Pacific Comprehensive Energy Partnership, which has identified $6 billion in U.S. export credit and government financing to promote clean energy development in the Asia-Pacific region.

Looking ahead, we will target these and other resources towards greater penetration of renewables in the global energy mix on both a small and large scale, including through our participation in the Sustainable Energy for All Initiative and accelerating the commercialization of renewable mini-grids. These efforts include:

- **Natural Gas.** Burning natural gas is about one-half as carbon-intensive as coal, which can make it a critical "bridge fuel" for many countries as the world transitions to even cleaner sources of energy. Toward that end, the Obama Administration is partnering with states and private companies to exchange lessons learned with our international partners on responsible development of natural gas resources. We have launched the Unconventional Gas Technical Engagement Program to share best practices on issues such as water management, methane emissions, air quality, permitting, contracting, and pricing to help increase global gas supplies and facilitate development of the associated infrastructure that brings them to market. Going forward, we will promote fuel-switching from coal to gas for electricity production and encourage the development of a global market for gas. Since heavy-duty vehicles are expected to account for 40 percent of increased oil use through 2030, we will encourage the adoption of heavy duty natural gas vehicles as well.

- **Nuclear Power.** The United States will continue to promote the safe and secure use of nuclear power worldwide through a variety of bilateral and multilateral engagements. For example, the U.S. Nuclear Regulatory Commission advises international partners on safety and regulatory best practices, and the Department of Energy works with international partners on research and development, nuclear waste and storage, training, regulations, quality control, and comprehensive fuel leasing options. Going forward, we will expand these efforts to promote nuclear energy generation consistent with maximizing safety and nonproliferation goals.

- **Clean Coal.** The United States works with China, India, and other countries that currently rely heavily on coal for power generation to advance the development and deployment of clean coal technologies. In addition, the U.S. leads the Carbon Sequestration Leadership Forum, which engages 23 other countries and economies on carbon capture and sequestration technologies. Going forward, we will continue to use these bilateral and multilateral efforts to promote clean coal technologies.

- **Energy Efficiency.** The Obama Administration has aggressively promoted energy efficiency through the Clean Energy Ministerial and key bilateral programs. The cost-effective opportunities are enormous: The Ministerial's Super-Efficient Equipment and Appliance Deployment Initiative and its Global Superior Energy Performance Partnership are helping to accelerate the global adoption of standards and practices that would cut energy waste equivalent to more than 650 mid-size power plants by 2030. We will work to expand these efforts focusing on several critical areas, including: improving building efficiency, reducing energy consumption at water and wastewater treatment facilities, and expanding global appliance standards.

Negotiating Global Free Trade in Environmental Goods and Services: The U.S. will work with trading partners to launch negotiations at the World Trade Organization towards global free trade in environmental goods, including clean energy technologies such as solar, wind, hydro and geothermal. The U.S. will build on the consensus it recently forged among the 21 Asia-Pacific Economic Cooperation (APEC) economies in this area. In 2011, APEC economies agreed to reduce tariffs to 5 percent or less by 2015 on a negotiated list of 54 environmental goods. The

APEC list will serve as a foundation for a global agreement in the WTO, with participating countries expanding the scope by adding products of interest. Over the next year, we will work towards securing participation of countries which account for 90 percent of global trade in environmental goods, representing roughly $481 billion in annual environmental goods trade. We will also work in the Trade in Services Agreement negotiations towards achieving free trade in environmental services.

Phasing Out Subsidies that Encourage Wasteful Consumption of Fossil Fuels: The International Energy Agency estimates that the phase-out of fossil fuel subsidies – which amount to more than $500 billion annually – would lead to a 10 percent reduction in greenhouse gas emissions below business as usual by 2050. At the 2009 G-20 meeting in Pittsburgh, the United States successfully advocated for a commitment to phase out these subsidies, and we have since won similar commitments in other fora such as APEC. President Obama is calling for the elimination of U.S. fossil fuel tax subsidies in his Fiscal Year (FY) 2014 budget, and we will continue to collaborate with partners around the world toward this goal.

Leading Global Sector Public Financing Towards Cleaner Energy: Under this Administration, the United States has successfully mobilized billions of dollars for clean energy investments in developing countries, helping to accelerate their transition to a green, low-carbon economy. Building on these successes, the President calls for an end to U.S. government support for public financing of new coal plants overseas, except for (a) the most efficient coal technology available in the world's poorest countries in cases where no other economically feasible alternative exists, or (b) facilities deploying carbon capture and sequestration technologies. As part of this new commitment, we will work actively to secure the agreement of other countries and the multilateral development banks to adopt similar policies as soon as possible.

Strengthening Global Resilience to Climate Change: Failing to prepare adequately for the impacts of climate change that can no longer be avoided will put millions of people at risk, jeopardizing important development gains, and increasing the security risks that stem from climate change. That is why the Obama Administration has made historic investments in bolstering the capacity of countries to respond to climate-change risks. Going forward, we will continue to:

- Strengthen government and local community planning and response capacities, such as by increasing water storage and water use efficiency to cope with the increased variability in water supply

- Develop innovative financial risk management tools such as index insurance to help smallholder farmers and pastoralists manage risk associated with changing rainfall patterns and drought

- Distribute drought-resistant seeds and promote management practices that increase farmers' ability to cope with climate impacts.

Mobilizing Climate Finance: International climate finance is an important tool in our efforts to promote low-emissions, climate-resilient development. We have fulfilled our joint developed country commitment from the Copenhagen Accord to provide approximately $30 billion of climate assistance to developing countries over FY 2010-FY 2012. The United States contributed approximately $7.5 billion to this effort over the three year period. Going forward, we will seek to build on this progress as well as focus our efforts on combining our public resources with smart policies to mobilize much larger flows of private investment in low-emissions and climate resilient infrastructure.

II. *Leading Efforts to Address Climate Change through International Negotiations*

The United States has made historic progress in the international climate negotiations during the past four years. At the Copenhagen Conference of the United Nations Framework Convention on Climate Change (UNFCCC) in 2009, President Obama and other world leaders agreed for the first time that all major countries, whether developed or developing, would implement targets or actions to limit greenhouse emissions, and do so under a new regime of international transparency. And in 2011, at the year-end climate meeting in Durban, we achieved another breakthrough: Countries agreed to negotiate a new agreement by the end of 2015 that would have equal legal force and be applicable to all countries in the period after 2020. This was an important step beyond the previous legal agreement, the Kyoto Protocol, whose core obligations applied to developed countries, not to China, India, Brazil or other emerging countries. The 2015 climate conference is slated to play a critical role in defining a post-2020 trajectory. We will be seeking an agreement that is ambitious, inclusive and flexible. It needs to be ambitious to meet the scale of the challenge facing us. It needs to be inclusive because there is no way to meet that challenge unless all countries step up and play their part. And it needs to be flexible because there are many differently situated parties with their own needs and imperatives, and those differences will have to be accommodated in smart, practical ways.

At the same time as we work toward this outcome in the UNFCCC context, we are making progress in a variety of other important negotiations as well. At the Montreal Protocol, we are leading efforts in support of an amendment that would phase down HFCs; at the International Maritime Organization, we have agreed to and are now implementing the first-ever sector-wide, internationally applicable energy efficiency standards; and at the International Civil Aviation Organization, we have ambitious aspirational emissions and energy efficiency targets and are working towards agreement to develop a comprehensive global approach.